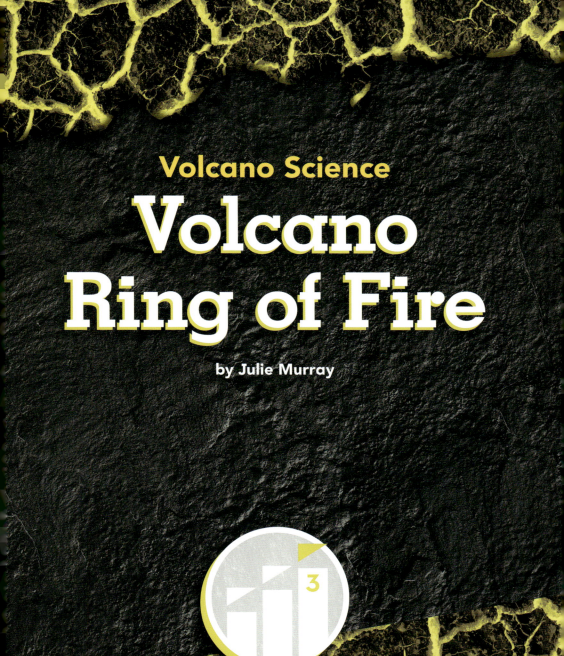

Volcano Science
Volcano Ring of Fire

by Julie Murray

Level 1 – Beginning
Short and simple sentences with familiar words or patterns for children who are beginning to understand how letters and sounds go together.

Level 2 – Emerging
Longer words and sentences with more complex language patterns for readers who are practicing common words and letter sounds.

Level 3 – Transitional
More developed language and vocabulary for readers who are becoming more independent.

abdobooks.com

Published by Abdo Zoom, a division of ABDO, PO Box 398166, Minneapolis, Minnesota 55439. Copyright © 2023 by Abdo Consulting Group, Inc. International copyrights reserved in all countries. No part of this book may be reproduced in any form without written permission from the publisher. Dash!™ is a trademark and logo of Abdo Zoom.

Printed in the United States of America, North Mankato, Minnesota.
052022
092022

Photo Credits: Alamy, Getty Images, NASA, Shutterstock
Production Contributors: Kenny Abdo, Jennie Forsberg, Grace Hansen, John Hansen
Design Contributors: Candice Keimig, Neil Klinepier

Library of Congress Control Number: 2021950301

Publisher's Cataloging in Publication Data

Names: Murray, Julie, author.
Title: Volcano ring of fire / by Julie Murray.
Description: Minneapolis, Minnesota : Abdo Zoom, 2023 | Series: Volcano science | Includes online resources and index.
Identifiers: ISBN 9781098228439 (lib. bdg.) | ISBN 9781098229276 (ebook) | ISBN 9781098229696 (Read-to-Me ebook)
Subjects: LCSH: Volcanoes--Juvenile literature. | Geography--Juvenile literature. | Volcanism--Juvenile literature. | Physical geography--Juvenile literature.
Classification: DDC 551.21--dc23

Table of Contents

Volcano Ring of Fire 4

Moving Plates 8

Positive Outcomes 18

More Volcano Facts 22

Glossary 23

Index 24

Online Resources 24

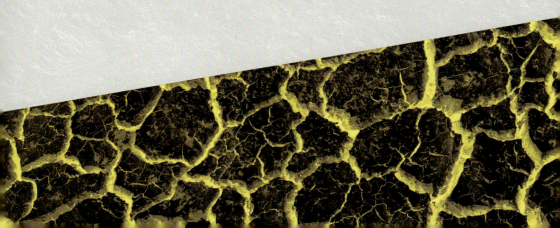

Volcano Ring of Fire

The Ring of Fire is a horseshoe-shaped belt made up of **plate** boundaries and volcanoes. It stretches 25,000 miles (40,234 km) along the rim of the Pacific Ocean.

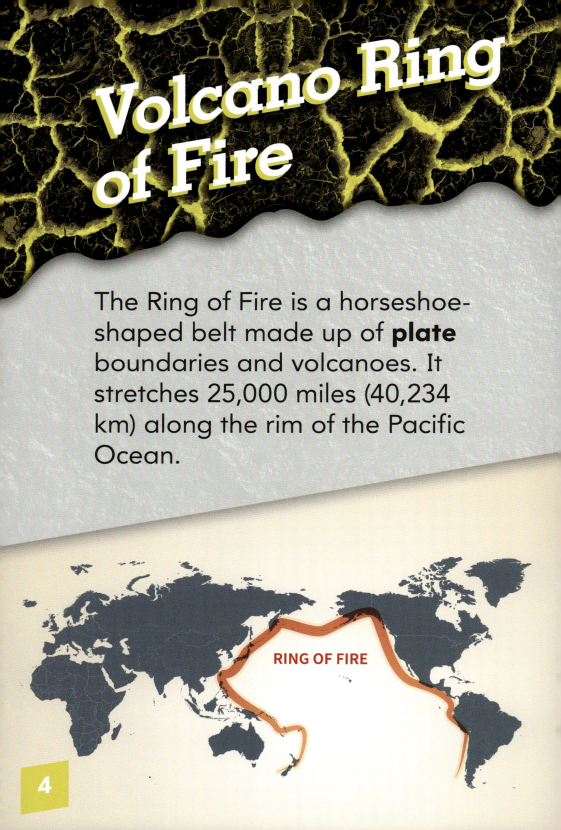

RING OF FIRE

More than 450 volcanoes are found in the Ring of Fire. That is 75% of all volcanoes on Earth!

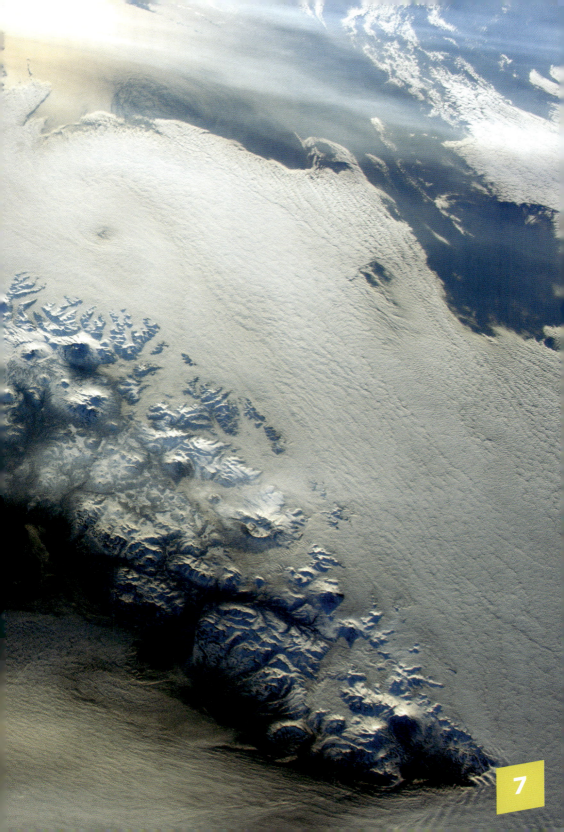

Moving Plates

The Earth's outer **crust** is made up of **plates**. The Ring of Fire is a meeting point for many of these plates. Volcanic activity is very common at these points.

TECTONIC PLATE

8

The **plates** move and collide with one another. This causes **pressure** to build up deep in the Earth.

The **pressure** pushes magma to the surface. This is what causes a volcano to form and erupt.

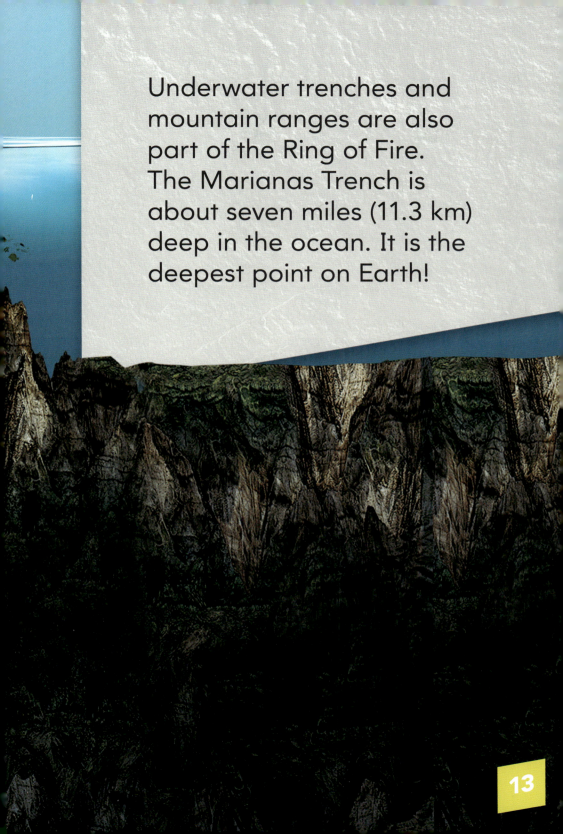

Underwater trenches and mountain ranges are also part of the Ring of Fire. The Marianas Trench is about seven miles (11.3 km) deep in the ocean. It is the deepest point on Earth!

The Ring of Fire runs along the western coast of North America. The Cascades are a mountain range in this area. There are 18 volcanoes in the range. Mount St. Helens is one of the most active.

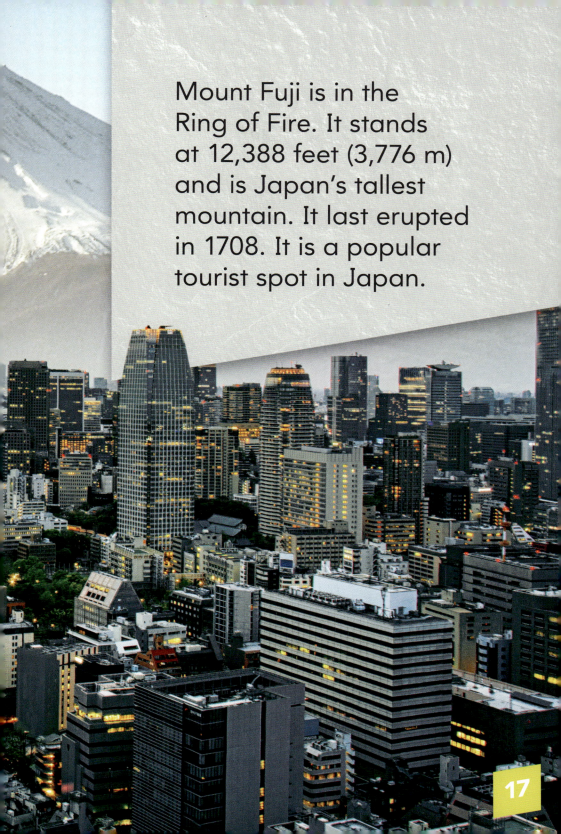

Mount Fuji is in the Ring of Fire. It stands at 12,388 feet (3,776 m) and is Japan's tallest mountain. It last erupted in 1708. It is a popular tourist spot in Japan.

Positive Outcomes

Being in the Ring of Fire can be a good thing too. Lava and ash put **nutrients** into the ground. The areas around volcanoes have some of the world's most **fertile** farmlands.

Geothermal energy comes from volcanic activity. Scientists use the heat that is trapped under the Earth's surface to produce energy. The Ring of Fire is one of the most active geothermal areas on Earth.

More Volcano Facts

- 90% of the world's earthquakes also occur in the Ring of Fire.

- 15 countries are in the Ring of Fire.

- Scientists believe that 80% of all volcanic eruptions occur underwater.

- Most of the world's largest eruptions within the past 11,700 years occurred in the Ring of Fire.

- Japan is home to 10% of the world's volcanoes.

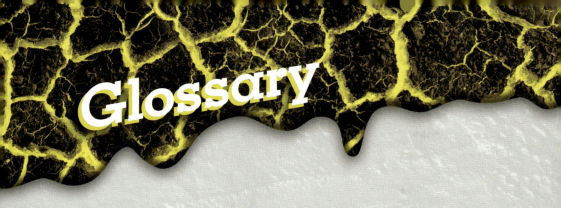

Glossary

crust – the outer layer of Earth.

fertile – producing or able to produce plant life.

geothermal energy – heat energy produced in the interior of earth, or energy derived from earth's heat.

nutrient – something that helps plants live and grow.

plate – also tectonic plate, one of the plate-like segments of the Earth's crust and upper mantle. The plates form the outer shell of the planet. They move very slowly. Where the plates meet each other (a boundary), their movement can cause or play a part in the eruption of volcanoes, the building of mountains, and earthquakes.

pressure – a steady force upon a surface.

Index

ash 19

Cascade Mountain Range 15

crust 8

eruptions 11, 15, 17, 19

geothermal energy 21

Japan 17

lava 19

length 4

magma 11

Marianas Trench 13

Mount Fuji 17

Mount Saint Helens 15

Pacific Ocean 4, 13

tectonic plates 8, 10

United States 15

Online Resources

Booklinks
NONFICTION NETWORK
FREE! ONLINE NONFICTION RESOURCES

To learn more about the Ring of Fire, please visit **abdobooklinks.com** or scan this QR code. These links are routinely monitored and updated to provide the most current information available.